Two Orphan Cubs

Barbara Brenner
and May Garelick

Illustrated by
Erika Kors

A TRUMPET CLUB SPECIAL EDITION

Published by The Trumpet Club

666 Fifth Avenue, New York, New York 10103

ISBN: 0-440-84255-7

Reprinted by arrangement with the Walker Publishing Company, Inc.
Printed in the United States of America
April 1990

10 9 8 7 6 5 4 3 2
UPC

Text design by Laurie McBarnette

Two Orphan Cubs

This is a true story about black bears.
It begins in winter
in a bear den hidden between two rocks . . .

Winter

The mother black bear was asleep.
Beside her lay her two tiny hairless bear babies—
each about the size of a hamster.
Their eyes weren't open yet.
They couldn't see anything in the den.
They couldn't even see their mother,
right there beside them.

But the cubs knew she was there.
They could feel her tongue licking them.
They could feel her soft furry body.

The two newborn bears nuzzled
the warm fur of their mother's belly
until they found her milk.
They nursed, purring like kittens.
Then they slept.
And then they nursed some more.
And they slept again.
The mother bear slept, too.
But she didn't eat or drink
all through the cold days of winter.

Two months went by.
Inside the den, the cubs were growing.
Now they were as big as rabbits.
Their bodies were covered
with black fur like their mother's.
It covered their ears, their legs
and their fat little bellies.
Their eyes were open now.
They could see.
And their baby teeth were coming in.

Spring

Winter was turning into spring.
Green grass began to poke through the snow.
Slivers of sun shone into the den.
Mother bear stirred, grunted, and went back to sleep.

Then one day she woke up.
What was it that woke her?
Outside the den the air was soft and warm.
Did she smell it?
Could she feel spring in her bones?

The mother bear sat up and blinked away sleep.
Something told her this was the day—
the day the cubs would have
their first look at green grass—
the day she would begin to teach them
to poke their paws into a log for bugs
and to scurry up a tree
to get away from danger.
But first she would go out alone and look around.

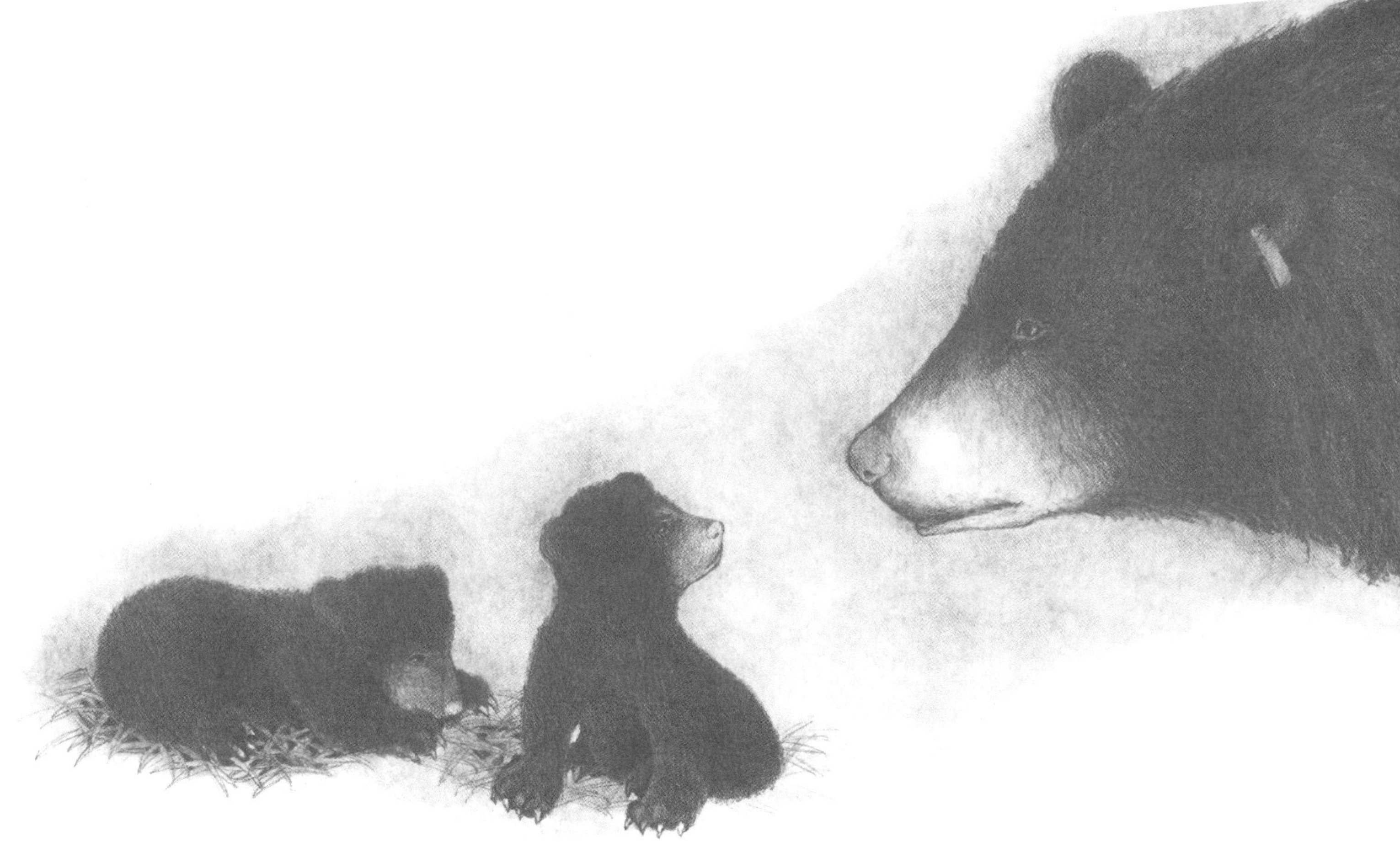

With soft, huffing noises
the mother called the cubs to her.
She licked them.
Then she woofed at them as if to say,
Stay.
At three months, the cubs didn't know much.
But they understood that sound.
They knew enough to stay.

The mother black bear left the den.

The two little bears waited for her to come back.

Hours passed.
The last rays of sun left the woods.
The den was damp and dark.
The cubs began to shiver.
It was cold without their mother's warm fur.
They snuggled close and nuzzled each other,
looking for milk.
But there was no milk.
They cried themselves to sleep.

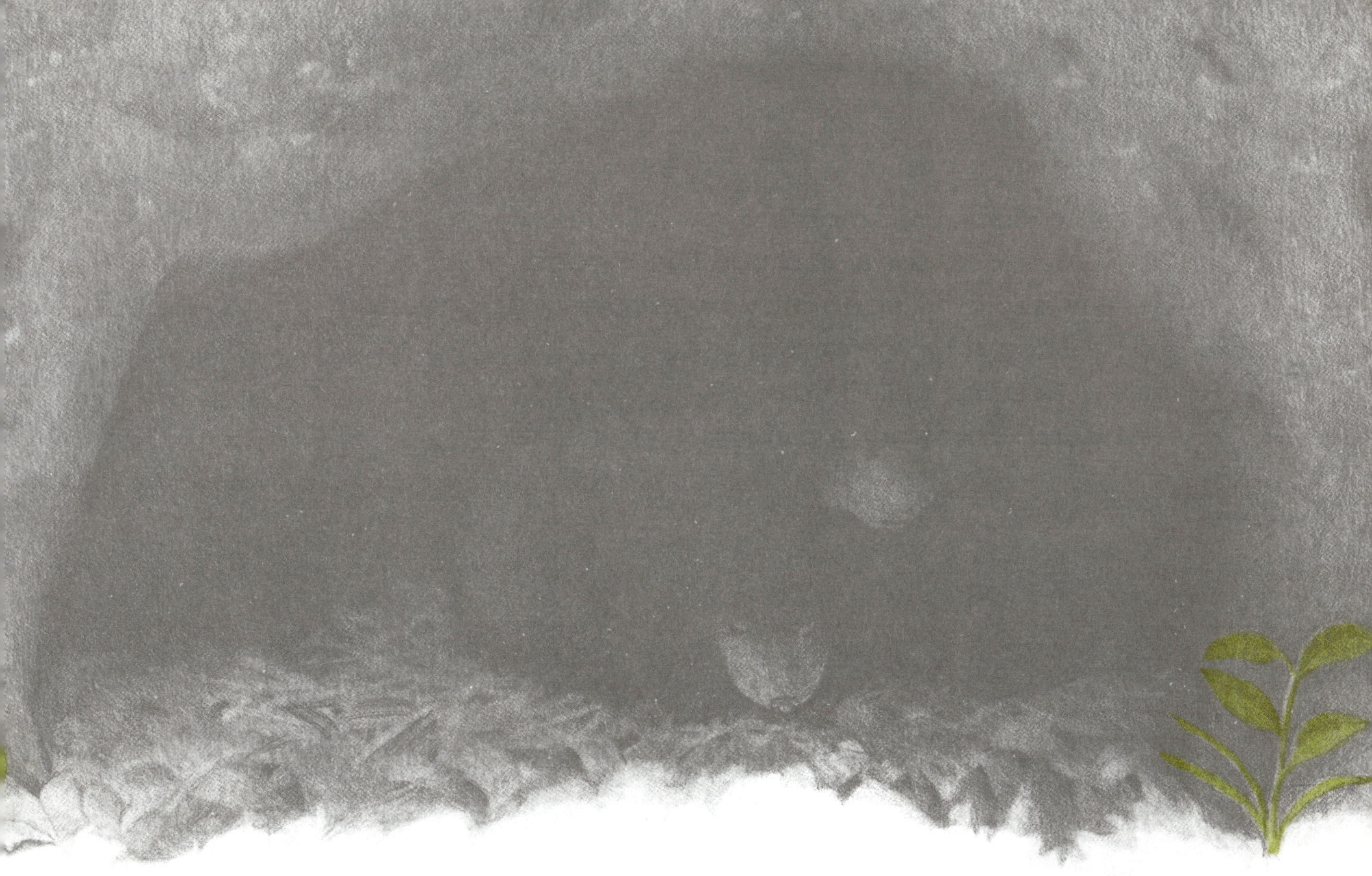

When the cubs woke up the mother bear was still not there.
Now they were very hungry.
Their yelps began to sound frantic.
How long could two nursing cubs last
without a mother?

At ten o'clock that morning,
Gary Alt's jeep came through the woods.
Gary is a wildlife scientist who studies black bears.
He had come to the den to check
on the bear he called Lulu.
He heard whimpering coming from inside.

When he shined his light into the den
he could see the eyes of the cubs.
But no mother. No Lulu.
Gary was worried.
A mother bear never leaves her cubs for long.
And she always comes back
if she hears them crying.
What could have happened?

Gary looked around outside.
Behind a clump of bushes,
he saw boot prints,
and two pink ear tags.
Lulu was the only bear with pink identification tags.
Gary had put them on her himself.
Then he saw the empty gun shell
and he knew what had happened.
Lulu had been shot by a poacher.

Gary felt terrible.
There was nothing he could do about Lulu.
But he had to try to save her babies.

He crawled into the den
and reached for the cubs.
When he grabbed them, they nipped at his hands.
But their baby teeth couldn't bite
through his thick gloves.

He tucked the trembling little bears into a sack
and crawled out of the den.
He talked softly to them.
"There, there. Easy does it.
We're on our way to find you
a new Mama."

Gary drove to another den deep in the woods.
In that den was the bear he had named Molly.
He knew that Molly was nursing two cubs of her own.
He hoped that she would adopt Lulu's cubs,
and make them part of her family.

But how to get them into her den—
that was tricky.

If Molly awoke and smelled Gary, she might attack him.
Even gentle black bears may attack
to protect their cubs.
But he had to take the chance.

Gary lifted the cubs out of the sack
and put them on a shovel.
The cubs squealed and squirmed,
but when he patted them they settled down.
He tiptoed to the den.
No sound from inside. Molly was asleep.
He slid the baby bears off the shovel
and into the den hole.

Did Molly see the two cubs?
Would she lick them—or bite them?
Would she adopt them?
Or would she push them out of the den?

Gary waited.
Ten minutes passed.
Then he heard licking sounds.
He took a chance
and turned on
his flashlight.

There was Molly *licking* the orphans.
They had found her milk
and she was letting them nurse.
She had adopted them and would
raise them as her own.
Now Molly had *four* cubs.
And the two orphans had a new mother.

Mother Molly

A week later Molly was wide awake.
She sniffed the air.
It smelled sweet and warm.
Something told her this was a good day
to take her cubs outdoors.
She looked out. It was safe.

With a gentle huff,
she called the four little bears.
She led them out of the den
and into the field.

Then she did what all bear mothers do.
She took the cubs to a clump of green.
They had their first taste of grass and skunk cabbage.
She showed them how to poke their paws
into a log for bugs.
She taught them to scurry up a tree
to get away from danger.
Soon Molly will show them the black bear trails.
She will teach them
where to find wild blueberries,
and where bees keep their honey.
In two years, when they know
what bears have to know,
the cubs will go off on their own.

Gary Alt is a real person. He's a wildlife biologist whose specialty is black bears. As part of his research, he tracks their movements and sets harmless bear traps to catch them. When Gary catches a bear he tranquilizes it and gives it a number tattoo and a pair of ear tags. If it's a female bear, he fits her with a radio collar so he can follow her movements. The collar sends out radio signals that he can pick up on a receiver. These signals help him to locate the bear at any time, and to check on her movements. If something should happen to her later, when she has cubs, he can locate another female bear with cubs and place the orphan cubs with her. In this picture, Gary is fitting a radio collar on a tranquilized black bear.

Gary Alt has placed over one hundred orphan cubs with new bear mothers. And he has tagged over two thousand bears.

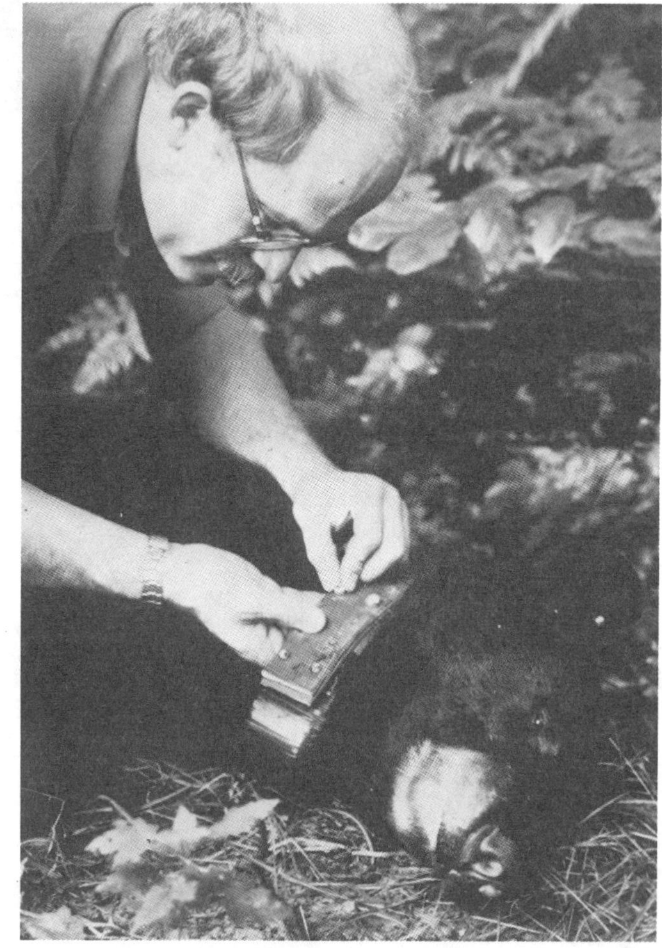